Discovery at Blue Moon Bay

by Sharon Kahkonen

Copyright © by Harcourt, Inc.

All rights reserved. No part of this publication may be reproduced or transmitted in any form or by any means, electronic or mechanical, including photocopy, recording, or any information storage and retrieval system, without permission in writing from the publisher.

Requests for permission to make copies of any part of the work should be addressed to School Permissions and Copyrights, Harcourt, Inc., 6277 Sea Harbor Drive, Orlando, Florida 32887-6777. Fax: 407-345-2418.

HARCOURT and the Harcourt Logo are trademarks of Harcourt, Inc., registered in the United States of America and/or other jurisdictions.

Printed in China

ISBN-13: 978-0-15-362467-4
ISBN-10: 0-15-362467-1

6 7 8 9 10   0940   16 15 14 13
4500452468

Visit *The Learning Site!*
www.harcourtschool.com

First Encounter

It was a perfect day for George and Mary's favorite activity—beachcombing! The pure white sand felt warm under their bare feet. The crystal clear water of the Caribbean Sea was a beautiful turquoise blue. George and Mary were looking for shells, bits of coral, and anything else that the waves may have washed up on shore. They thought that someday they would like to open a museum in their back yard for the other kids in the neighborhood, or maybe even for the tourists. They had found some pretty interesting things—crab shells, shark teeth, and sea shells of all different shapes and sizes.

But on this day they found something truly extraordinary. Mary was turning over a piece of coral in her hand, wondering what it looked like when it was alive, when she caught something out of the corner of her eye.

"George, look!" Mary shouted, as she pointed out to the sea.

2

Something big and black was coming toward them out of the water. Their first impulse was to run, but they were too curious. They stood very still, frozen in their tracks. It looked like the head of a sea creature, with one huge eye! Slowly, its shiny black body emerged from the water. Very clumsily, it waddled to the shore. Its huge webbed feet were obviously better suited for swimming than for walking. Once on shore, off came the mask, off came the tank, and off came the flippers. It wasn't a weird sea creature after all: it was a scuba diver!

"Hi there!" the scuba diver greeted them.

"Hi!" Mary replied. "What are you doing?"

"My name is Dr. Fonseca. I'm a marine biologist," the woman answered. "I'm studying the coral reef out there. We're trying to find out more about why local fishers aren't catching as many fish as they used to, and why the corals are fading. The reef and the economy of the entire island may be in trouble."

"Oh, yes, we know about the fish," George replied, with a concerned look. "My father is a fisher, and he has been catching fewer and smaller fish every year."

"So what do you think might be causing this?" asked Mary.

"It could be because of global warming, or pollution, or sediments washing into the ocean, or over fishing, or some combination of these factors. We still don't know for sure. Hopefully, our research will help shed some light on this mystery."

"What do you have in those sacks?" asked George.

"I've collected some specimens to take back to the laboratory." She pointed down the beach to a cement, two-story building. "It's down there. I'm studying the effects of pollution on their growth."

"Can we come with you? Can we help you in some way?" asked Mary.

Setting down her sacks, Dr. Fonseca thought for a moment. She was extremely busy, but Mary and George looked keenly interested. She realized that saving the reefs would require everyone's help. She needed to convince fishers, farmers, and people in the tourist business to do things differently. They needed to know how important it was to save the reefs and how they could help. And they needed to be educated about the biology of the reefs to understand how to take better care of them. Maybe some energetic young people could help. She could tell them about the reef. They could tell their parents. In turn, their parents could tell other people.

"How would you like to bring your whole class to the lab?" she asked.

"Do you mean it? That would be fantastic!" said Mary.

"Okay, let me see what I can arrange with your teacher. Just now, I have to get back to work. But I promise, we'll meet again soon."

"This is George, and I'm Mary. We go to Middletown School, and our teacher's name is Mr. White," Mary called after Dr. Fonseca, who was already making her way down the beach.

"I'll give him a call. See you soon, Mary and George!" she called back.

"Don't forget!" exclaimed George.

"I won't," Dr. Fonseca replied. "I promise."

At the Marine Biology Lab

Later that week, the students in George and Mary's class went to Dr. Fonseca's lab. There were rows and rows of tanks, filled with corals, sponges, and fish of all colors, shapes, and sizes.

"All of these things live out there in the reef?" asked George, eyes wide with wonder.

"Yes, and many more. There are thousands of different animals living out there. These are only some of the most common types."

"These animals don't look common to me. I've never seen some of them before. They're dazzling!" replied Mary.

Dr. Fonseca smiled. "You're right, Mary, coral reefs are dazzling. Nowhere else in nature has so much beauty and variety of life been crowded into so small an area. Every square foot holds new surprises."

"What are these things, Dr. Fonseca?" "Are they plants or animals?" "Can they move?" "How do they eat?" The students bombarded her with an endless string of questions.

"Hold on!" Dr. Fonseca laughed, happy to hear the children's interest. "Let's take this a step at a time. To answer one question that many of you have, yes, all of these creatures are animals, just like we are."

"They sure don't look like animals. Why don't they move like other animals?" asked George.

Dr. Fonseca began to lecture, "Corals and sponges are two phyla of very primitive animals. First, let's talk about all of the different kinds of animals that exist and how they are arranged into groups. You see, scientists have classified all animals into groups according to their traits. That is, animals with similar traits are placed together in the same group."

Alicia, a friend of Mary's, asked, "Since humans are animals, what group do we belong to?"

"Good question," Dr. Fonseca answered. "First of all, as you all know, humans belong to the animal kingdom. But kingdom is just the first level of classification. A kingdom includes the largest number of different organisms."

"Are there just two kingdoms—plants and animals?" asked George.

Dr. Fonseca explained, "Actually, many scientists use a five-kingdom classification system. It was once believed that all living things were either plants or animals. Then scientists created another kingdom for the fungi, including mushrooms and molds, because they are so different from plants. And as scientists learned more and more about different organisms, including microscopic one-celled organisms like bacteria, they decided that they needed more kingdoms for classification. In the future, as scientists learn more, they may need more than five kingdoms. But today we're only going to talk about the animal kingdom."

"Let's see whether you can keep track of the seven levels of classification," said Dr. Fonseca. "There are seven levels in the classification system. The highest level is the most general and contains the most organisms. Each level gets more specialized until you get to the lowest level, which contains only one organism. Starting at the top, or most general level, the levels are: Kingdom, phylum, class, order, family, genus, and species."

"How can you remember all those levels?" said George with a bit of surprise.

"I've had practice," said Dr. Fonseca. "But you can use tricks to help remember things like this. Some people use a sentence, 'King Phillip came over for great spaghetti.' The 'k' in *King* is for kingdom, the 'p' in *Phillip* is for phylum, and so on." She started to write. "Humans would be classified like this."

This is what Dr. Fonseca wrote:

Alicia looked puzzled. "That seems very complicated. Are all animals classified like this?"

Dr. Fonseca nodded, "Yes, all living things fit into this system of classification. It actually makes things a lot simpler. There are over a million different animals on Earth that scientists have identified. But they can all be grouped, or classified, according to their characteristics.

For example, humans belong to the chordate phylum. This is because they have a nerve cord in their back. Humans also belong to a smaller group of chordates—the subphylum Vertebrata. This is because they have a backbone. What other animals do you think belong to this subphylum?"

"I know! Dogs and cats have a backbone," answered George.

"So do birds and fish!" added Alicia.

"Exactly," replied Dr. Fonseca, "Many of the animals that are most familiar to us are vertebrates. The next level of classification is class. Humans belong to the class Mammalia. All mammals are warm-blooded. Most also give birth to their young and feed them with the mother's milk. Humans are more closely related to other mammals, like dogs and cats, than they are to some of the other vertebrates, like fish or frogs or birds."

"What about the animals in the ocean? Are we closely related to any of them?" asked Mary.

"Whales, dolphins, and porpoises are mammals. Of course, there are many fish that belong to the same phylum—chordata—as humans. But most of the animals that live in the ocean are much more primitive," said Dr. Fonseca.

"Why is that?" asked George.

Dr. Fonseca explained, "Life began in the sea. For millions of years, life was present only in the sea. Today there are still many primitive forms of animals living there. Come, let me show you some of them."

Dr. Fonseca took the students on a tour of the lab. She told them about seven different phyla of animals that live in coral reefs. From simplest to most complex, they are sponges, cnidarians, segmented worms, mollusks, echinoderms, arthropods, and chordates. As they walked through the lab the students learned about each phylum.

Sponges

Sponges are very primitive animals. Their body walls are two layers thick and are held up by fibers that act like a skeleton. When sponges die, they leave behind their fibrous skeletons. In the center of the sponge is a space called the central cavity. The walls contain pores that lead into the central cavity. In the central cavity, the pores are surrounded by flagella—tiny whips that are constantly moving. As they move, they draw water and pieces of food into the sponge, where the food is digested.

Cnidarians

This phylum contains some of the most beautiful of all animals—the corals and jellyfish. Some cnidarians have a body shaped like a hollow sac that has one opening. This type of cnidarian is called a polyp. The opening serves as a mouth. Surrounding the opening are stinging tentacles. Polyps use their tentacles to push food down into their mouths. Coral formations are made by millions of tiny coral polyps living together in colonies.

As they grow, the animals cement themselves in place by depositing limestone around the lower half of their bodies. As new coral animals grow, the limestone formation that they are building also grows. Corals may look like branching trees, large domes, or even organ pipes. The living coral animals form a cover over the limestone in beautiful shades of yellow, orange, purple, and green.

Segmented Worms

Earthworms are segmented worms that live in the soil. Bristleworms and fireworms are segmented worms that live in coral reefs. The segmented worms are much more complex than sponges and corals. They have many tissues that are organized into organs and organ systems. They have a head and a tail, a complete digestive system, nerves, and blood vessels.

Mollusks

The word *mollusk* comes from a Latin word that means soft. The mollusks have very soft bodies. All mollusks except octopods and squids have hard shells to protect their soft bodies. Clams, scallops, and mussels are in a class called bivalves because their shells have two sides, or valves. A bivalve can open its shell and stretch out its foot. It can also stretch out a special food tube. Water containing food particles is sucked through the tube into the bivalve's gills. The gills filter out food particles. Snails belong to a different class because they have one shell. Squids and octopods belong to the class Cephalopoda, meaning head-footed. They have tentacles that extend from their heads. Squids and octopods can move quickly by squirting a jet of water from their bodies.

Echinoderms

Starfish, sea urchins, and sand dollars are echinoderms, meaning "spiny-skinned." The echinoderms have well-developed digestive systems and simple nervous systems. Starfish are the best-known echinoderms. The undersides of their arms are covered with tube feet, which act like tiny suction cups. A starfish can wrap its arms around a clam, using its tube feet to hold on. Then the starfish's stomach stretches out of its mouth and enters the clam. It digests the body of the clam and then sucks the digested material into its own body.

Arthropods

Lobsters and crabs are examples of arthropods that live in the sea. They belong to the class Crustacea. Insects, which belong to the class Insecta, are examples of arthropods that live on land. All arthropods have bodies made of several parts and hard external skeletons. They also have jointed feet, which is what the name *arthropod* means.

Chordates

Fishes are vertebrates, which means that they have skeletons and backbones. Sharks, rays, and skates are fish that have skeletons made of cartilage. They belong to the class Chondrichthyes. Parts of our own bodies, like the end of the nose and the ears, are made of cartilage. Most cartilage is softer than bone. Bony fish have skeletons made of bone tissue, like our own skeletons. They belong to the class Osteichthyes. A wide variety of brilliantly colored, bony fish live in coral reefs. They have many specialized sense organs. They can see, hear, sense pressure changes, and feel vibrations. They also have complex digestive and circulatory systems. Reef fish are specially adapted to live among corals. The corals provide endless nooks and crannies in which they can hide. The fish are as brilliantly colored as the corals themselves. In open waters they would be very conspicuous, but among the corals, they are much less noticeable.

Tom, one of the more mischievous students, tapped on the side of a tank. An octopus quickly retreated into a tiny hole. "Who's in there?" he asked.

"Read the label," said Dr. Fonseca. "Look here, It says *Octopus briareus.* That's its scientific name. Its common name is Caribbean reef octopus. It's a very interesting animal that is rarely seen in the daytime and is one of the great hunters of the night. It's highly intelligent and capable of rapid changes in both color and skin texture. It's a master of disguise and can slip through the smallest of openings."

"Why don't people just call it by its common name, instead of Octopus bra-ra-whatever? Wouldn't that be a lot simpler?" asked Tom.

Dr. Fonseca explained, "The common name may seem a lot simpler, but the fact is, this octopus has many different common names. In Germany, it's called 'Pulpo' or 'Krahe.' In France, it's called 'Poulpe ris.' In Spain, it's called 'Pulpo de arrecife.' But no matter where you are in the world, it has the same scientific name—*Octopus briareus.* Using the scientific name makes things less confusing. When I use the name *Octopus briareus,* other scientists

know exactly what animal I'm talking about. Incidentally, the scientific name is always underlined or in italics so that people know that it is a scientific name when they read it."

"So how do scientists come up with scientific names?" asked Alicia.

"The scientific name is actually made up of the last two levels of classification—in this case, the genus name *Octopus* and the species name *briareus*," explained Dr. Fonseca. "Who remembers the other levels of the classification system?"

"'King Phillip came over for great spaghetti.' *King* stands for kingdom; *Phillip* is for phylum; *came* is for class; *over* is for order, *for* is for family; and the last two levels—*great* is for genus; and *spaghetti* is for species," said Tom, beaming.

Dr. Fonseca laughed, "Absolutely! So here's how the Caribbean reef octopus is classified."

Dr. Fonseca wrote:

"I still don't understand how scientists decide which group the Caribbean reef octopus belongs to at each of those classification levels," said Mary, scratching her head.

"Scientists have been observing animals, including mollusks, for a long time. They look at a number of traits, such as body structures, and come up with a system for grouping them into the seven levels of the classification system. Actually, they sometimes put other sublevels between the seven main levels, such as subphylum or subspecies. At each level, they place animals with similar traits in the same groups. They give a name to each group and describe its features. This ordering helps them study and identify the millions of animals in the world. This whole process is called taxonomy," explained Dr. Fonseca.

"What happens if someone discovers a new species that no one has ever seen before?" asked George.

"Well, George, do you remember the day I met you and Mary on the beach? That was a very special day for me—not only because I met you two, but also because I found a type of bivalve that I had never seen before. I think I may have discovered a new species!" exclaimed Dr. Fonseca.

"How can you know for sure?" asked George.

"We're trying to identify it using a key to the bivalves," said Dr. Fonseca. "It is very closely related to bivalves in the *Barbatia* genus, but it has traits that are very different from those of any other bivalve. If it turns out to be a new species, I get to name it! Does anyone have any good ideas?"

"Name it after Blue Moon Bay, since that's where you found it," said Tom.

"Hmmm," said Dr. Fonseca. "*Barbatia bluemoonensis.*"

"No, no. I have a better idea," said Mary. "It should be named after Dr. Fonseca!"

"*Barbatia fonsecae?*" said Dr. Fonseca.

"That's a great idea!" said George. "All in favor, say 'Aye'!"

"AYE!" shouted everyone.